C000038628

Pioneering
Spirits

Pioneering Spirits

The 12th Rolex Awards for Enterprise

Contents

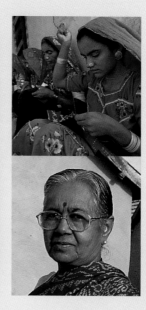

Associate Laureates 76

A World of Enterprise

Arctic Circle

Tropic of Cancer

Pacific Ocean

Atlantic Ocean

Equator

Pacific Ocean

Zenón GOMEL APAZA
Peru

Tropic of Capricorn

Cristian DONOSO
Chile

Antarctic Circle

Arctic Ocean

Alexandra LAVRILLIER
Siberia

Rory WILSON
nited Kingdom

Julien MEYER
Spain

Shafqat HUSSAIN
Pakistan

Pacific Ocean

Chanda SHROFF
India

Runa KHAN MARRE
Bangladesh

Pilai POONSWAD
Thailand

Atlantic Ocean

Indian Ocean

Brad NORMAN
Australia

Antarctica

Foreword

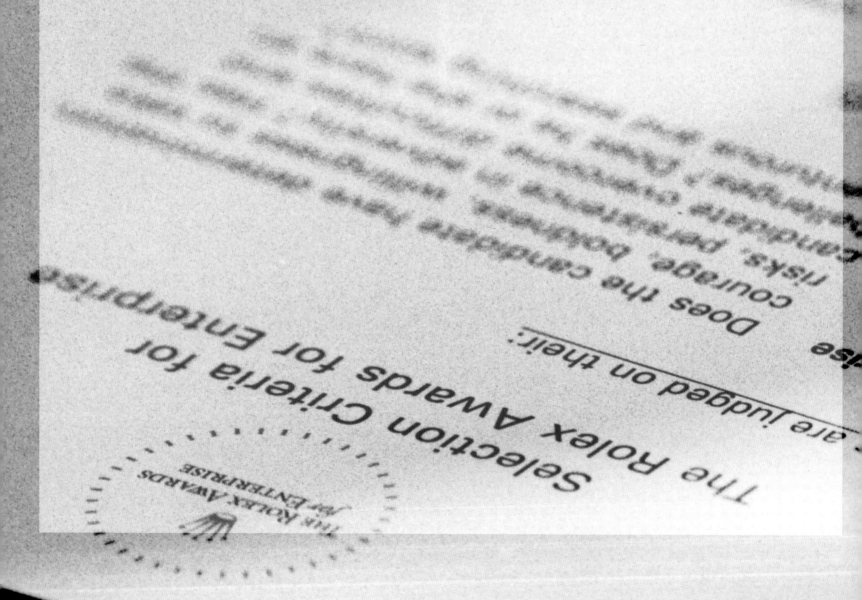

The Rolex Awards
for ENTERPRISE

Selection Criteria for
The Rolex Awards for Enterprise

Does the candidate have, or are judged on their:

courage, boldness,
risks, persistence,
candidate

big adrenalin rush. I never felt frightened, but I did keep my arms down and made myself small.

'Even with something as big as a whale shark, you're not afraid – and nor is it. It is a calming experience. You feel at one.' Swimming alongside its head, Norman has seen its little eye turn, observing him – perhaps a glimmer of acknowledgement. 'Maybe it just thinks I'm a big remora [sucker fish],' he laughs. Nonetheless, he respects the shark's brute power, and has assisted in the drafting of guidelines for divers and tour operators worldwide explaining how to behave around whale sharks.

Norman's love of the ocean was born on the golden beaches of Perth, on Australia's Indian Ocean coastline, where he body-surfed as a youngster. This led to diving and, via a science degree, to a deep interest in marine conservation which he has pursued as a researcher and fisheries management consultant.

His encounter with the Ningaloo whale sharks was a life-altering experience. The shark was an unknown, and there was little money for its study or conservation. Norman survived hand-to-mouth on sporadic grants, and funded much research himself. Burning the midnight

Mimicking a feeding whale shark, a giant sieve net captures zooplankton in the rich waters of Ningaloo Reef. Norman and a colleague collect the whale food, some organisms as small as one millimetre, for study. Unlike other filter-feeders, the whale shark does not rely only on forward motion to draw plankton over its gill-rakers, but can also hang vertically in the water and 'suck' food.

oil, he mounted national and international campaigns for the whale shark's conservation, emerging as a global authority on the animal and its needs. He helped authorities develop plans for its protection and wrote scientific reports and information for visitors.

Many mysteries are yet to be resolved. While young male sharks gather at Ningaloo, no one knows where the females collect or where the sharks breed. The key to studying their thin, dispersed and cryptic demographics lay in identifying individuals. Norman's painstaking research proved each has a pattern of white spots on its body as distinctive as a human fingerprint. This gave him the idea of using underwater camera images as a practical, non-invasive way to identify individuals. In 1999 Norman set up the ECOCEAN Whale Shark Photo-identification Library on the Internet, a global project to record sightings and images.

Despite the growing body of information, Norman had no efficient way to compare shots of whale sharks taken from different angles, under varying conditions and fish postures. In 2002, American computer engineer and fellow diver Jason Holmberg contacted him. After discussion, Holmberg agreed to help organize and automate the ECOCEAN database. He explained the photo-ID problem to a friend, NASA-affiliated astronomer Zaven Arzoumanian, whose colleague Gijs Nelemans brought to their attention a technique used by Hubble Space Telescope scientists for mapping star patterns, known as the Groth algorithm, which the team then adapted to map the patterns of white spots on the shark's hide. It took many months of intense mathematical calculations and computer programming to refine the algorithm for use on a living creature – but in the end they gained a breakthrough for biology, a reliable way to identify individuals in virtually any spotted animal population. In December 2005, the three described their findings in the *Journal of Applied Ecology*. More than 500 whale sharks have been identified and added to the database using the technique.

For survival, whale sharks depend on huge bursts of tiny sea life which reflect the condition of the oceans and their bio-productivity. Since whale sharks travel immense distances to collect food, the demographics of these fish can be an indicator of ocean health – and of the human impact on it.

Divers worldwide can now follow Norman's simple guidelines for photographing whale sharks and log their images, activities and locations on the ECOCEAN site, where they are automatically catalogued, matched and, if possible, identified as belonging to a known individual. Each new image will help Norman compile a map of where whale sharks live and their migratory patterns. And contributors receive

**The protected shallows of Ningaloo Marine Park are one of
the few places where whale sharks appear in large numbers
at about the same time every year – April, May and June –
and where divers can easily photograph them (overleaf).
Norman saw his first whale shark here in 1995 and decided to
devote his life to conserving this largest of fish.**

notification of all past and further sightings of the sharks they photograph. Together, the images are helping to build a global picture of the
abundance, health and fluctuations of the whale shark population.
'Just about anyone with a disposable underwater camera can now
play a part in helping to conserve whale sharks, and so monitoring the
health of the oceans,' Norman explains. 'It gives people a direct stake
in its stewardship.'

With the Rolex Award money, Brad Norman is devoting two years
full-time to his project, training local authorities, tourism operators
and 20 research assistants around the Pacific, Atlantic and Indian
oceans to observe, record and protect whale sharks. In this way he will
develop whale shark photography as a significant tool for conservation.

He will also explain to those who hunt the shark that there is more
to be gained by leaving it alive. Ningaloo's whale sharks draw more
than 5,000 visitors a year, mainly from April to June, generating ecotourism worth an estimated US$10 million. A live whale shark earns
far more than a dead one. 'The whale shark is worth saving – and we
can do something about it,' says Norman. 'It is a big, beautiful and
charismatic animal, and not dangerous. It is a perfect flagship for the
health of the oceans.'

Nomad School Teaches New and Ancient Skills

By Pierre-Yves Frei

In south-eastern Siberia, a nomadic people are trying to preserve their way of life against the march of modern society. The traditional culture of the Evenk, who excel at reindeer herding, hunting and fishing, has been eroded through contact with Western civilization. For eight years Alexandra Lavrillier, a brilliant French ethnologist, has been fighting alongside them to save their heritage, setting up a nomadic school that will give Evenk children the chance to receive a modern education without having to sacrifice their ancestral traditions.

Alexandra Lavrillier France

Alexandra Lavrillier is French, but she has always felt she belonged in the far North. 'As a child, I spent hours in the Musée de l'Homme [Museum of Mankind] in Paris, gazing at the displays about the Inuit and the peoples of Siberia,' she says. Now she spends much of the year far from the comforts of Paris, in a region far more austere and demanding: in the middle of the Siberian taiga, where winter temperatures can drop to -50°C.

In 1994, after studying Russian at the Institut national des Langues et Civilisations Orientales (National Institute of Eastern Languages and Civilizations) and the Centre d'Etudes mongoles et sibériennes (Centre for Mongolian and Siberian Studies) in Paris, and learning Yakut, an indigenous Siberian language, from immigrants living in France, the young Frenchwoman accompanied an expedition organized by several French photographers. For three months she travelled, as the expedition's interpreter and ethnologist, the length and breadth of Yakutia, in Siberia, the region that had fascinated her as a child, and there she had a life-changing encounter with the Evenk people. 'I found them the most welcoming of all the Siberian people, and they had conserved their culture and their language better than anyone else,' she explains. Twelve years later, her passion is unabated. Now a highly respected ethnologist, 36-year-old Lavrillier is married to an Evenk nomad from the Stanovoy Mountains. Together they have a little daughter.

The first known mention of the Evenk, as hunters and reindeer herders, dates back to the 17th century when the Russian empire, in constant expansion eastwards, came into contact with the ethnic group. At the time, there was no respect for cultural diversity, and, as the centuries passed, there was little improvement in the status of the Evenk. In the 20th century, their animist rituals, in which shamanism plays an important role, were frowned upon by the Soviet government. The Evenk needed a great deal of determination and courage to preserve their beliefs and traditions, often in the utmost secrecy. To add to the difficulties, the authorities decided late in the 1960s that Evenk children should follow the normal school curriculum, even if this meant that for months at a time, year after

Her heart belongs in the Siberian taiga, home to the nomadic Evenk people. Laureate Alexandra Lavrillier was fascinated by the region as a child and now spends much of the year with the reindeer herders. She has founded a travelling school, housed in a tent, where (preceding pages) an Evenk child trudging to school through new-fallen snow can learn both the Russian curriculum and Evenk skills and rituals.

'Alexandra Lavrillier's knowledge of the local language, techniques, institutions, rituals and culture is exceptional. She also proved extremely brave in the difficult conditions of life in the taiga and extremely diplomatic in the delicate post-Soviet context.'

Roberte Hamayon director, Centre d'Etudes mongoles et sibériennes, Paris

Modern technology and nomad ways are taught by Lavrillier and two Evenk teachers who joined the school upon earning their teaching certificates. Computers will help the youngsters cope with the modern world, while learning how to clean a bear skin or create a cap of grey reindeer fur relate to their daily lives. In warm weather classes are held outside (overleaf).

year, the children had to attend state boarding schools far from their families and nomadic life.

Alexandra Lavrillier knew that to safeguard the future of Evenk culture, children's education was paramount. For eight long years she has devoted her time and energy to setting up the travelling school that the Evenk nomads had dreamed of for so long. The nomad school has been up and running since the start of 2006, after a successful campaign by the French ethnologist to overcome the many administrative obstacles in her path. Finally, she succeeded in doing what no one else had done for a nomadic school for a Siberian minority: after she won approval from the Russian education ministry and the relevant authorities in the region of Amur, in south-eastern Siberia, the school was granted the status of 'official experimental school', recognition that may pave the way for similar experiments elsewhere in Siberia.

Now at last Evenk children have a school that can travel with them, that is adapted to their lifestyle and, most importantly, does not require them to be separated from their parents for long periods so they can attend classes. The Rolex Award presented to Lavrillier will pay for at least the crucial first three years, covering the cost of teaching materials, multimedia equipment, a team of reindeer for transport and salaries for three Evenk teachers, including Lavrillier, and for a guide. The Award will also cover the costs of printing and distributing books that the nomadic school will produce on Evenk language and culture.

The 23 six- to ten-year-olds from several different camps now attending the Evenk nomadic school have all the benefits of a full educational programme. The teachers travel from one camp to another, with the time spent at each camp depending on the educational level and needs of the children, who then continue their work on their own until the school returns. As well as the traditional Russian curriculum, the subjects include English, French and an Internet-awareness module, using a computer powered by an electric generator. Lavrillier insisted on language and computer courses

'This courageous Frenchwoman has set up a school that travels
to nomadic children in remote Siberia to teach them about
their traditional culture, along with modern languages and
computer technology. Her project is a major step forward for
the Evenk people.'

Motoko Ishii president of Motoko Ishii Lighting Design

Learning the ways of their elders and recording them for
future generations, Evenk childen have busy school schedules.
A pupil photographs a wooden sledge for a book about Evenk
culture that Lavrillier intends to publish. Early education
includes how to lasso and ride a reindeer, an animal vital to
their survival. Lavrillier leads youngsters in a traditional dance.

because she believes the students must have the tools they will need to deal with the modern world – and to benefit from it. 'The state infrastructures that used to provide jobs for Evenks are being shut down one after another,' Lavrillier says. 'So, in the very near future, they will need to be ready to defend their rights and learn about the market economy. Some of them might even want to start their own small businesses.'

Education for the modern world is now balanced by the cultivation of the Evenk heritage in the travelling school, which allows the children to stay with their parents and elders and continue to be part of their own community. The young Evenk can learn to fish, look after reindeer and be initiated into various rituals. They will also have an opportunity to study their traditions in class – and a chance to help conserve them, as Lavrillier is relying on their cooperation for the books the school is producing, including a guide to the flora and fauna of the taiga that will explain how the Evenk use and manage this environment with its often extreme conditions. The guide will also explain how the Evenk separate and spread out in winter in order to make the most of the few resources available. When warmer weather returns, they attend a large gathering before accompanying their herds of reindeer to high-altitude pastures where the summer heat is less harsh on the animals. The guide on the flora and fauna will be followed by a handbook on the Evenk language and a book on their traditions and beliefs. The guidebooks should help the many Evenk people who have taken up residence in various parts of Russia to rediscover their roots. For Lavrillier, the fact that the people concerned have access to the information gathered about them is all-important.

These new guides will be critical to revitalizing Evenk culture. Since *perestroika* in the late 1980s, Russia has rediscovered the richness of its ethnic peoples and has tried to rehabilitate them. The Evenk, like about 30 other Siberian minorities, have been granted special status and a degree of autonomy intended to enhance their identity, beliefs and traditions. Resources, however, are often too

When temperatures rise, small streams spring up in the taiga, signalling time to move camp. The Evenk will seek higher altitudes and change location about every two weeks, moving an average 1,000 kilometres every year. Not all children attend Lavrillier's school. Some, like these two Evenk boys beating a rhythm with reindeer leg bones, leave their families for boarding school in town.

scarce to repair the damage done by centuries of cultural erosion. There are only 30,000 Evenk left in Russia. Most of them have completely abandoned the nomadic way of life and the taiga, and now live in villages and towns, with no memory of their ancestral hunting and fishing techniques. More often than not, these men and women have switched to agriculture, and two-thirds of them can no longer speak their traditional language. Generally there is little to envy in the way of life of these settled Evenk. They find it hard to fit into modern Russian society, and, as with all the Siberian minorities, the proportion of unemployed Evenk is far higher than the Russian average. Only a few of them reach higher education.

However, it is among these settled Evenk that Lavrillier has recruited the two teachers who work with her at the nomadic school, although it took her considerable energy to find them and persuade them to accept the jobs. To leave modern living, however modest, in order to return to nomadic life – especially one which requires enduring the harsh Siberian winters – is a difficult decision to take, even when jobs are scarce.

Alexandra Lavrillier's successful achievements on behalf of the Evenk rely on her outstanding knowledge of their world, gained partly through study, but mainly in the field. While she has carried out research in libraries and archives in Siberia, the principal source of her understanding are the many months spent with these people – almost every year since her first trip in 1994 – sharing their nomadic life, studying their traditions and lifestyle, on the roads and in the forest, in winter and in summer, inside a tent or riding reindeer. Lavrillier is now making the Evenk way of life better known to the outside world, through lectures and articles, and, in 2005, a doctoral thesis presented to the Ecole des Hautes Etudes in Paris for which she received the unanimous congratulations of the jury. 'It is only because I know all about the Evenk nomads, their living conditions and their environment, that I had the confidence to put forward this project for a nomadic school, as a feasible, viable project,' says Lavrillier.

In the broad expanse of the taiga, the Evenk travel at will, usually in small groups. A group portrait includes Lavrillier, her husband and daughter and two other families who journey together. They are often joined by the two teachers who travel between camps, leaving the children to study on their own between visits. One youngster shows the reindeer fur cap she made. Lavrillier says those 'who remain nomadic are the sole custodians of this culture'.

Revive Brilliant Embroidery in a Severe Land

By Alexa Schoof Marketos

In a remote part of India, one woman has established a movement to revive a local form of artistic expression, hand embroidery, creating a sustainable means of income. The region of Kutch once had a long and rich tradition of embroidery which made a welcome contrast to the region's austere landscape. But, from the 1960s onwards, synthetic materials and machine work pushed this craft close to extinction. Acutely aware of its cultural, social and spiritual value, Chanda Shroff is preserving this unique heritage while promoting an exquisite art form and empowering women in highly conservative societies.

Chanda Shroff India

The painstaking and beautiful craft of hand embroidery dates back several thousand years. One of its traditional homes is Kutch, a corner of the Indian state of Gujarat. Known for its intricate and diverse styles, Kutchi embroidery has, since the 1960s, suffered a decline due to a modern emphasis on speed and profit, and a growing reliance on machinery and synthetic fabrics. An Indian woman, Chanda Shroff, aged 73, has worked tirelessly and voluntarily for almost four decades to reverse this decline.

Determined that the traditional techniques of Kutchi embroidery will be handed down to future generations of women, Shroff will use the funding from the Rolex Award to create a mobile resource centre to promote the embroidery. This constitutes the second phase of a two-phase project, aptly called 'Pride and Enterprise', which has its roots back in 1969 when Shroff set up Shrujan – Sanskrit for 'creativity' – a non-profit organization dedicated to helping the drought-afflicted communities of Kutch. Virtually an island, Kutch is bordered to the south by the Arabian Sea and to the north by immense salt deserts. Descendants of immigrants and invaders, its 1.2 million people represent a highly diverse range of ethnic groups and cultures. Yet all these cultures share a rich tradition of embroidery.

For the people of Kutch, embroidery is more than just decoration for household goods: it is an important means of personal, social and spiritual expression. Each piece of intricate embroidery brings creativity and beauty into daily life, providing a welcome foil to the harsh climate and austere landscape. Traditionally, embroidered articles formed an integral part of a girl's dowry, while for royals and nobles these articles were symbols of status and wealth.

Today each ethnic group and community retains its own distinctive motifs and lexicon of stitches, handed down through the generations in 16 distinct styles of embroidery. While many of the stitches are universal, the craftswomen create unique combinations with a great degree of complexity. Rabari embroidery, for example, is vigorous, with bold shapes and designs taken from mythology and inspired by the desert surroundings. Ahir embroidery is, by contrast, curvilinear in style, animated with motifs such as peacocks, parrots,

Dedicated to the art of embroidery, Chanda Shroff is both restoring and changing old ways in the region of Kutch in the Indian state of Gujarat. Under her guidance, women are gathering together to create exquisite handiwork despite differences in religion and status – and they are building their own economic base without giving up their traditional roles.

Busy hands stitch bright patterns in one village, while a contemporary design of butterflies and flowers was produced in another. Throughout the region, Shroff's self-help groups pursue their individual style of embroidery. Parmaben, a woman of the Ahir people and one of the first to join Shroff's project, shows the Laureate how she executes a design.

Festival finery worn by a young embroiderer (overleaf) shimmers with mirror fragments. The intricate panel portrays the traditional tree of life of the Ahir community.

scorpions, elephants and flowers. Soof embroidery, on the other hand, is a counted thread style which uses a single stitch to create highly geometrical designs. Other styles use mirrors or a form of quilting, and colour selection also differs: Rabari embroidery features earth tones and white, while Ahir embroidery is characterized by dark violet, gold and red.

Shroff, who has a teaching diploma in crafts, started 38 years ago by providing 30 women from one village with raw materials and assistance with designs. With the establishment of Shrujan, women who had never had occasion to mix began working together; over time they found common ground, initially in the sharing of embroidery techniques and designs, and later in shared personal experience. Today Shrujan, based near Bhuj, the capital of Kutch, has directly benefited more than 22,000 women from 120 villages and all castes across Kutch. 'Little shoots of inter-caste acceptance have begun to sprout,' Shroff explains. 'Just a few years ago, Rajiben, a master craftswoman from the Dalit community [previously considered to be "untouchables"] would not have been allowed to step into the homes of the higher-caste women of the Ahir and Sodha communities. Today, after a painful struggle on both sides, Rajiben is accepted by them as their teacher. The women all sit and work together in their homes, exchanging ideas and even food.'

A cornerstone of Shroff's vision has been an unswerving commitment to the quality that is central to the Kutchi embroidery tradition, despite the conditions in which many of the people live. 'I was deeply shaken by the plight of the Kutchi people and especially the women,' she says of her initial encounters with them. 'Here were a people reduced to utter helplessness and dependency, even while they possessed in their hands and minds skills such as few others could claim.' Rejecting the modern preference for synthetic materials, the craftswomen primarily use silk and cotton to create high-quality products for fashion and decoration. Each craftswoman is encouraged to stitch her name into each piece of embroidery, and, in doing so, her

role as artist and guardian of a unique cultural heritage is reinforced. The national recognition they now receive and the income from the sale of the embroidery have brought them deep respect in their communities. The steady flow of revenue from outside customers whom Shroff has found to buy the products is slowly uplifting the status of women, allowing them to invest in land, pay for health care and improve their families' nutrition levels.

Pride and Enterprise was conceived when Shroff realized that Shrujan would be short-lived if she could not inspire younger craftswomen to recognize the richness of this ancient craft: 'I needed a big idea, an idea at the intersection of conservation, education, enterprise and empowerment; an idea that could light a fire, especially in the hearts of the younger generation.'

The first phase, completed in 2004, was the creation of 1,200 hand-embroidered display panels representing the different styles and carefully stored in the basement of Shrujan's headquarters. About 600 rural craftswomen took part in the work, creating 90-centimetre by 120-centimetre display panels, each designing between one and four panels. The artists included 85-year-old Parma Balasara, one of the first craftswomen of Shrujan, and about 400 women under 30, with the older artisans mentoring the younger ones.

Each panel took between three months and a year to complete, depending on the complexity of the design, with some incorporating long-forgotten motifs. This resource is a celebration of the skills of the craftswomen, a mirror within which they can see themselves as custodians of an artistic heritage. Richard Franklin, a former head of design at the Smithsonian Institution's Museum of Asian Art in Washington D.C., said of the panels: 'These are works of great artistry, and the collection is a breathtaking testament to the aesthetics and vision of the artisans who created them and the tradition they embody.'

The mobile resource centre that Shroff is setting up will take selected panels to the craftswomen, many of whom are not permitted to leave their villages. Trained facilitators will accompany the unit, and videos, photographs and monographs will be prepared for each embroidery style, including demonstrations on how to execute the

Shrujan means 'creativity' in Sanskrit, and Shroff's non-profit organization lives up to its name. Promoting a renaissance in embroidery and marketing the products bring concrete results to village artists. Shrujan employees visit villages to pay for finished work and provide materials for new creations. A young woman proudly shows Shroff her work.

'What struck me most about this project was how it springs from the needs and reality of the women involved. Thousands of women are contributing their exquisite work to this project which is in turn allowing them to inspire one another.'

Denise Bradley vice chancellor of the University of South Australia

designs, explanations of the natural and cultural influences that inform these styles, and biographies of the craftswomen.

A preliminary collection of 50 panels has already been taken to nine villages. 'This seemingly ordinary act has had a dramatic – almost explosive – impact on the village communities,' Shroff explains. 'Exhibiting the panels led the women, both young and old, to look at themselves and their skills in an entirely new way. That Kutchi embroidery could be so rich and diverse in expression, that such exquisite work is possible in present times, that women like themselves could produce such high-quality work – this has been a revelation to the villagers.'

Shroff is also organizing self-help groups to train the craftswomen to gradually assume the roles of designers, saleswomen, entrepreneurs and teachers. To date, there are 19 self-help groups made up of 380 women. 'I am convinced of the need to develop systems that will eventually allow for the decentralization from the mother organization Shrujan,' says Shroff. 'I would be most happy if there were no longer only one Shrujan, but instead many mini-Shrujans all over Kutch.' Informal craft schools also feature in Shroff's plans to overcome the social isolation of the women and to stimulate innovation in the craft. Groups of 15 young women attend the schools for a three-month cycle.

Shroff, who was chosen as a Laureate of the Rolex Awards for her plan to ensure the survival of an exquisite art form in a way that creates a sustainable source of income for the women of Kutch, recognizes that her vision is an ambitious one. But, having spent more than

Seeking to encourage young women to take up the skills of their elders, Shroff asked those in her broad network to produce panels showing different designs and styles, such as the bold geometrics of the Rabari (above). The result: 1,200 works of art, many of which travel by truck from village to village where Shroff explains the project, while artisans eagerly study the panels looking for inspiration and new ideas.

'The collection of panels is a breathtaking testament to the aesthetics and vision of the artisans who created them and the tradition they embody.'

Richard Franklin former head of design at the Smithsonian Institution's Museum of Asian Art

half her life working with the craftswomen, she speaks confidently of what they can achieve. 'The women of Shrujan are like my own family. We have been through so much together – a war, cyclones, droughts and, most recently, the earthquake [in 2001]. We have learned from one another and always we have found solutions together.' She is determined to make Kutch once again a rich source of traditional embroidery, to bequeath a legacy that will survive for thousands of years to come, a magnificent art form that provides, in her words, 'a support system for home-based women, as well as a reminder of the creativity and potential inherent in all women'.

Daring to dream and working to make the dreams come true, Chanda Shroff has inspired more than 22,000 women in 120 villages to value their talents and recognize their potential. Here she strolls through Hodko, a village rebuilt after the massive 2001 earthquake, where many women are now turning their lives around and ensuring the survival of an ancient art form.

Mobile Technology Records Daily Diaries of Animals

By Graeme O'Neill

English zoologist Rory Wilson is renowned for developing ingenious ways to track wild animals and record their behaviour without directly observing them. His latest invention, a lightweight electronic logging device, can go where satellite-based tracking devices cannot, to observe animals living in the wild. Wilson's new logger harnesses the laws of physics to accurately estimate the energy expenditure of animals on land, at sea and in the air. He is using this precious data to revolutionize research into the lives of threatened species in order to save them and their habitats.

♛ Rory Wilson United Kingdom

Rory Wilson's passion for penguins was born in 1962 when, as a four-year-old, he visited the zoo with his mother. Entranced, he watched as the black-and-white birds vanished below the surface of their murky pool, and bobbed up seconds later, many metres away. The submarine antics that so delighted the young boy would frustrate his efforts many years later to study his favourite birds in the wild. Armed with a master's degree in zoology from Oxford in the early 1980s, Wilson headed to the University of Cape Town to begin a Ph.D. study of the behavioural ecology of the African penguin (*Spheniscus demersus*). Africa's only endemic penguin was in rapid decline from a population of almost two million a century ago to only 220,000 in 1982 (and 180,000 today). The species is currently classified by the World Conservation Union (IUCN) as 'vulnerable to extinction'.

Standing on the shore of Marcus Island off the South African coast, on the first day of his field studies, Wilson watched the birds waddle into the sea to hunt small fish in the frigid, foggy waters of the Benguela Current. Dismay suddenly tempered his delight at the prospect of spending several years in the company of his favourite birds. How was he going to observe the behaviour of a species that spends much of its day submerged in the sea? His own seasickness and the amount of time the penguins spend under water soon convinced Wilson of the futility of trying to follow them in a small boat.

Wilson knew that his research could not advance unless he found a way to observe the bird under water. Lacking sophisticated tracking equipment, he created a simple, ingenious device to record, for the first time, how fast penguins swim. It comprised a harmless pellet of radioactive phosphorus embedded in a polystyrene bung and attached to a spring inside the barrel of a syringe. The bung moved as the water pressure varied with the penguin's swimming speed, creating a pattern of light and dark bands on a radiation-sensitive film strip stuck on the tube. Wilson's 'penguin speedometer', which was attached to the animal's breast with a small leather harness, cost him less than a dollar and allowed him to analyse the bird's swimming patterns.

This basic invention 25 years ago sparked an entire career as a researcher and inventor of innovative devices for monitoring wild

Intrigued with animals that move at night, swim the depths and hunt in remote places, for 25 years Laureate Rory Wilson has been creating miniature devices that record their movements and how they live, essential information for their survival. An emperor penguin (preceding pages) wears a recorder that will reveal the bird's strategies for acquiring food to feed its chick in the extraordinary Antarctic environment.

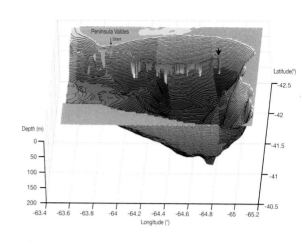

animals. Now aged 49, Wilson, who is professor of aquatic biology at the Institute of Environmental Sustainability at the University of Wales, has invented 24 novel devices or techniques which have enabled him and other scientists to study wild animals and to record their natural behaviour in new ways. The devices, three of which are sold commercially, can be used for studies lasting only hours or up to a year. Because they use little battery power and their data are stored on board, there is no need for bulky satellite- or radio-telemetry devices to be attached to the animals. His lightweight, miniaturized loggers cause minimal restrictions to animals' movements and behaviour.

Since the 1980s, huge advances in the electronics industry have helped many, not least Rory Wilson, produce a wide variety of small loggers. But ironically, as Wilson points out, this is complicating the monitoring of animal behaviour. 'There are now devices that sense a plethora of things: speed, acoustics, heart rate, heat loss, feeding, diving, and so on,' Wilson says. 'All this tends to drive biologists to search out a particular system to answer highly specific questions related to their own select species. Given the enormous numbers of animals in need of study in the world, we need to be consolidating powerful, cross-species logger systems, not all diversifying into our own tiny species corners.' Wilson adds that, according to the 2006 IUCN Red List, one in eight bird species and one in four mammals are classified as endangered.

Rory Wilson believes that the greatest need is for a single system that will determine the behaviour, energy expenditure and location of animals of many different species. Although Wilson's solution is technically complex, it is so simple in concept that he calls it his 'silly idea', comparing it to having animals 'keep a diary: an automatic diary that measures and records acceleration and heading'. Until recently he had insufficient funds to work on this device. Reviewers from funding institutions rejected his grant applications because they fell into unfamiliar terrain between biology and physics, or lacked specific hypotheses about the species on which he wanted to test the logger. His Rolex

Creatures that feed at sea pose a challenge for scientists who wish to monitor their activities. Where do they hunt, how deep do they dive? Wilson's logging device records the diving activity (diagram) of a Magellanic penguin off Argentina's Valdes Peninsula. The birds feed predominantly on fish that also appeal to man. If man's harvesting diminishes the fish stocks, the logger would indicate how this impacts on the penguins. The device has been successfully attached to lemon sharks in the warm Caribbean and to Weddell seals that dive in the dark, cold depths of Antarctic seas. The recorder is proving pivotal in understanding them in their hidden habitat.

'The science in this project is fascinating. A device like this will enable such a fantastic variety of behavioural studies to be carried out, we will get to learn about the physical, daily activities of animals in extraordinary detail.'

Mark Shuttleworth chairman of the Shuttleworth Foundation

Living underground by day and foraging at night, the badger is a model species to test Wilson's device. Despite intense study in Oxfordshire since the 1970s, scientists have been unable to get full insight into the badger's world. Translating the data from the logger into a diagram would reveal the route, pace and behaviour of such a wild animal, in effect creating a 'daily diary' of its activity.

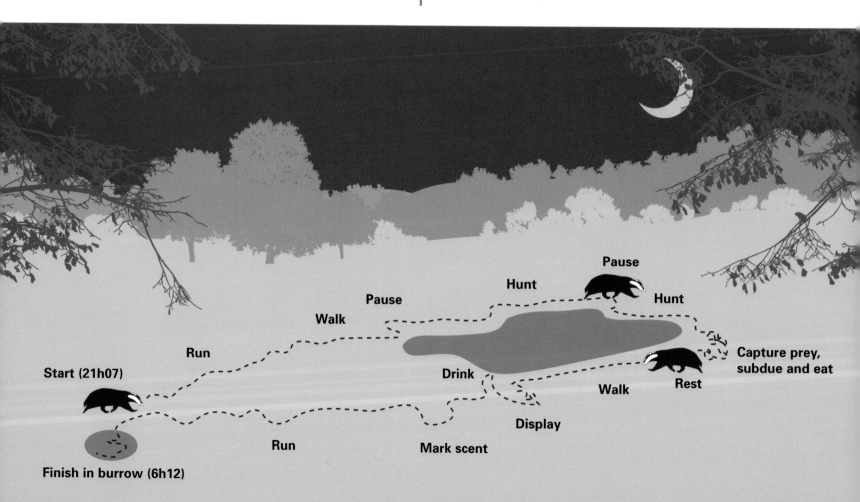

Pause

Hunt

Pause

Walk

Hunt

Run

Capture prey, subdue and eat

Start (21h07)

Drink

Walk

Rest

Display

Run

Mark scent

Finish in burrow (6h12)

Award will now provide the support needed to develop and test his revolutionary invention.

Wilson likens the logger, which weighs only 30 to 48 grams, to the black-box flight recorder that monitors changes in an aircraft's speed, altitude and heading. It contains a triaxial accelerometer, a tiny electronic device that monitors changes in an animal's acceleration. For example, a bounding kangaroo may be moving forwards, up/down, and sideways simultaneously. The accelerometer measures along all three axes up to 32 times a second, and, combined with a compass, the device determines the animal's speed, direction and position. Wilson's black box can do many things that widely used GPS (Global Positioning System) transceivers cannot, such as functioning in a dense forest, underground or in the ocean.

All animals expend energy staying warm, digesting food, and maintaining other vital body functions like breathing and pumping blood – but movement requires energy expenditure ten times higher. 'An animal that's not expending energy is dead,' Wilson says. Animals burn glucose to generate energy, consuming oxygen in the process, so by measuring an animal's oxygen intake in a sealed chamber called a

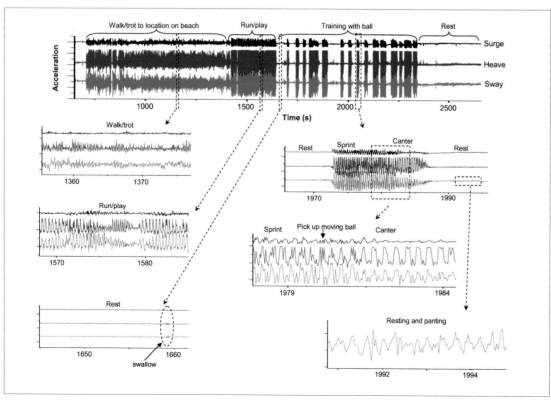

'Any one of Rory Wilson's devices would be a significant milestone for a single career, whereas he has made many. He has had more creative insights than a combination of us could ever hope to achieve.'

Dan Costa professor of ecology and evolutionary biology at the University of California, Santa Cruz

Wilson's best test subject, his Border collie, Moon, wears a logging device that records the dog's acceleration: forward (black), up and down (blue) and side to side (red). Correlating the energy needed for such activities as running and playing gives Wilson and his associates information that relates to dietary needs. An animal that exerts more energy than it takes in will be an animal in trouble.

respirometer, scientists can estimate how much energy it consumes just staying alive and warm, and how much it requires while walking, running or swimming. Wilson and his colleagues have already used the logger to record energy expenditure in wild cormorants, and were thrilled when their data corresponded to the figure predicted from trials determining the average oxygen consumption of five great cormorants tested in a respirometer. Zoologists will be able to use Wilson's black box to estimate the energy an animal expends flying, swimming, hunting, digging, feeding, fighting or mating. Adding these figures to the baseline energy needed to stay alive and warm will yield a reliable estimate of the species' total energy expenditure.

This information will have revolutionary consequences. By measuring the energy content of a species' natural diet, zoologists will know how much time a carnivore must spend hunting, or how long a herbivore must graze, to keep up its strength, grow and successfully reproduce – the ultimate aim of the game of life. 'A successful animal,' says Wilson, 'is one that takes in a lot more energy than it expends. Many conservation issues involve animals that are expending too much energy. Energy for an animal is like money for a human, but if an animal overdraws its budget, it dies. We haven't had a way of measuring energy expenditure in wild animals before.' Data about animals' energy expenditure will help conservationists understand what constitutes poor, average or optimum habitat, or what minimum area is required for a natural reserve to sustain a viable population of each species. The logger could help resolve important conservation questions, such as whether climate change, predation or over-fishing in its hunting grounds is responsible for the decline of the African penguin.

Wilson and fellow zoologists are testing the black box on species in Argentina, including imperial cormorants and armadillos. The device has already been trialled on beavers in Norway and the badgers of Wytham Woods in Oxfordshire, England. And closer to home, Wilson has done extensive tests of his logger on the family pet, a Border collie named Moon.

The importance of Wilson's device is exemplified by the insights that will be gained on the Oxfordshire badgers, which are of interest to those studying the evolution of social behaviour. David Macdonald, of Oxford University's Wildlife Conservation Research Unit, has been observing these badgers since the 1970s, making them amongst the world's most intensively studied carnivores, but he has always had difficulty tracking their detailed movements at night and observing their behaviour underground. 'Key to the issue,' he says, 'is the detail of where the badgers forage and where they scent mark, and Rory Wilson's amazing invention will reveal both.' Professor Macdonald adds: 'This information will not only help us understand the evolution of the badgers' mysterious social life, but will also be relevant to public health officials who need to understand their role in the transmission of bovine tuberculosis in cattle. The data we will gather in collaboration with Rory Wilson will therefore be not only interesting, but also practically useful.'

Wilson hopes his device will unlock many of the secrets of animal behaviour. Not only will it help save animals facing extinction now, it will also provide valuable data on many species almost certain to be threatened in the future. The beneficiaries of his project are, he says, 'the unthinkable number of animals that need to be properly understood now, tomorrow and in 20 years' time.'

The marvel of miniaturization allows Wilson to produce a device that fits in the palm of his hand and that Moon and other animals can wear without hampering their movements. Each device contains a compass that must be perfectly calibrated to provide reliable data. Wilson is developing his logger so it will provide a generic diary that will help scientists understand how wild animals work and what is important to them in order to make good conservation decisions.

Relaunch the Ancient Craft of Wooden Boats

By Fiona McWilliam

In recent decades the boat-building tradition of Bangladesh, which dates back several thousand years, has practically disappeared as all but the smallest boats have been superseded by noisy, diesel-powered, steel-plated vessels. Now Runa Khan Marre is restoring traditional, wooden boats and preserving the skills needed to construct them at a 'living museum', bringing glory – and tourists – to her country's waterways.

Runa Khan Marre Bangladesh

Until 20 years ago visitors to Bangladesh could witness a spectacle from another era, as hundreds of thousands of wooden sailing boats transported people and goods along the country's network of rivers. Since the mid-1980s, however, Bangladesh's riverscapes have changed beyond recognition: diesel-powered steel vessels have replaced nearly all traditional boats.

But 47-year-old Runa Khan Marre is determined that the Bangladeshi tradition of boat-making will not be lost. She and her husband, Frenchman Yves Marre, have ensured that Bangladeshis and tourists will soon be able to watch over 40 different types of traditional wooden river and sea craft being both restored and built from scratch by skilled craftsmen – and sail in them – at a 'living museum' on the banks of the River Dhaleswari, 20 kilometres north of the capital, Dhaka. With her Rolex Award, Runa Khan Marre will finalize preparation of the museum, pay for more boats to be restored, give employment to many craftsmen and bring pride and tourism revenues to her nation.

Bangladesh has 600 named rivers, totalling 24,000 kilometres in length, in an area a quarter the size of France; three major rivers, the Padma (Ganga), Jamuna (Brahmaputra) and Meghna, form the world's biggest delta. With between 750,000 and a million boats plying its waterways, Bangladesh is believed to have the world's biggest fluvial fleet. On average the nation's 147,570 square-kilometre land mass lies 5 metres above sea level, and during the monsoon period up to 60 per cent of the country is flooded. It is thus not surprising that powered boats, whose hulls cost a fifth the price of wooden ones, have been so successful, especially given that diesel engines overcome the navigational problems faced by traditional vessels with sails unable to stand up to strong winds. But the cost of modernization is the loss of the traditional fleet together with the skills needed to build and maintain wooden boats: most of the craftsmen are now aged over 50 in a country where life expectancy is 62. Their remarkable techniques, passed down verbally from generation to generation, date back more than 3,000 years to the Phoenicians.

Entrepreneur and humanitarian, Runa Khan Marre prizes her *malar*, the first wooden sailing boat that she and her husband, Frenchman Yves Marre, restored. The disappearance of such boats from her riverine country dismayed the Associate Laureate, who has found skilled craftsmen to revive one of the oldest traditions of Bangladesh.

 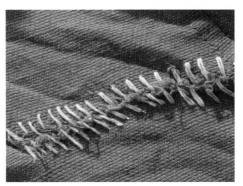 

The old ways still work at a shipyard outside Dhaka where Khan Marre is creating a 'living museum' of wooden boats. Workers cling to bamboo scaffolding to repair a *palowary*, a cargo vessel rescued from 35 years under water. Teams include blacksmiths to forge nails and staples, and weavers and tailors to make sails. Today metal boats (overleaf) have replaced the wooden craft that once served communities along the extensive waterways.

Leading a project to preserve boat-building is an unlikely occupation for a woman from an aristocratic Muslim background. But, early in her adult life, Khan Marre demonstrated her resourcefulness by setting up a fashion house and a security firm. Then, after overseeing the implementation of a major educational programme, she established, in 1998, the Friendship Association to provide health care in a floating hospital, flood relief and educational assistance to the impoverished inhabitants of the islands of the Brahmaputra River.

She was already known for her ability to implement challenging projects when, in 1994, she met Yves Marre, who was staying with her parents in Dhaka after sailing, from France to Bangladesh, a 38-metre river barge to be used for humanitarian purposes. He brought more than romance into her life as his passion for boats proved to be contagious. 'I discovered a new world,' Khan Marre recalls, 'and within months I was hooked.' One of their first joint achievements was the restoration of a *malar*, a 30-metre sailing boat they bought in 1996, which took local craftsmen over a year to bring back to life. She explains that her husband's technical expertise, coupled with her own ability 'to get things done', helped them establish a bond of mutual trust with the marginalized riverine people traditionally involved in boat-building.

'The technical skills and know-how required for the construction of these wooden boats and the rich heritage that the craftsmen produced over the centuries are irreplaceable treasures.'

Tommy Koh ambassador-at-large for the government of Singapore

Model boats, meticulously built to specifications, will be housed at the Living Museum. Funds from visitors, including the sale of models, will help support the project. Cruising near Dhaka, Khan Marre points to the wreck of a *patham*, now one of the rarest boats, but once plying the Meghna River in great numbers. Dexterous fingers weave a traditional canopy on a miniature *patham* that mimics that of a full-sized construction.

The couple then set up Contic River Cruise, which runs up-market river excursions on the *malar*. Established initially to repay the money they had borrowed for the restoration, the business now attracts influential foreign clients vital to Bangladesh's fledgling tourist industry. In 1999, determined to prevent boat-building skills from disappearing, Khan Marre searched Bangladesh for master ships carpenters, commissioning them to build scale models, each about 65 centimetres long, of boats from across the country. These replicas – there are now hundreds of them, reproducing 27 different types of boat – are built using the same techniques and materials as those for full-size boats. They provide an accurate record from which carpenters are able to build life-size boats.

'Once we saw the first models, and the success they enjoyed, we realized we had to do more,' Khan Marre recalls – and the idea of a living museum was born. Since 2004, carpenters, blacksmiths, ropemakers and sailmakers have been working at the Living Museum of Traditional Country Boats of Bengal, which opens to the public in April 2007. Carpenters from the Brahmaputra River have restored one of only two remaining 15-metre-long *palowary* boats, which have stapled hulls, while their counterparts from the Meghna River have constructed from scratch what is now the world's only *patham*, a fine example of a smooth-skinned boat. A team of carpenters from an island in the Bay of Bengal are building a seafaring *shampan* using

'These river boats deserve…urgent attention and development. The skills of the boat-building artisans are disappearing, and must be saved.'

Annie Montigny research director, Muséum national d'Histoire Naturelle, Paris

Matching ancient traditions with modern comforts, Khan Marre's *malar*, the *B613*, prepares to set sail for a sunset cruise where the Ganga and Brahmaputra rivers meet. The largest sailing boat in Bangladesh offers cruises to visitors who want to experience at first hand the cultural heritage that Khan Marre is preserving for the future.

techniques forgotten in Bangladesh, but revived with the help of Western marine architects and ethnologists, as well as museum documents and oral history. For each vessel, naval architects are documenting every stage of the boat-building, and their records will be made available to marine archives worldwide.

The project has given these people back their dignity, says Khan Marre, 'and the pride that comes from having great skills'. A local businessman has pledged to finance the construction of several buildings at the museum, including an exhibition area, a model-building workshop, shop and research centre. Khan Marre's project is making a vital contribution to her country's heritage. Annie Montigny, of the Muséum national d'Histoire Naturelle, in Paris, says that of all of Bangladesh's cultural heritage, 'these river boats deserve, more than anything else at present, urgent attention and development. The skills of the boat-building artisans are disappearing, and must be saved.'

Return to Ancient Ways Produces Crop Diversity

By Paul Jeffrey

Peruvian agronomist Zenón Porfidio Gomel Apaza is convinced that modern agricultural methods and technology are reducing biodiversity, depleting the soil and undermining community life, particularly in harsh regions like the Andes, where he grew up. He is using ancient skills and know-how from the region to ensure food security in rural communities and guarantee future generations access to their rich natural and cultural heritage.

Zenón Porfidio Gomel Apaza Peru

Zenón Porfidio Gomel Apaza thought he knew all about farming in 1994 when he packed up his books and returned from his university studies in agronomy to his village in the Peruvian Andes, 60 kilometres north of Lake Titicaca, where his ancestors had tilled the fields for generations. Yet in the harsh Altiplano, almost 4,000 metres above sea level, he realized that the modern agricultural methods he had studied so diligently had often produced a legacy of failed crops, depleted soils and dysfunctional communities. 'That was a turning point for me. My professional education didn't match the reality of the Altiplano,' Gomel Apaza says. 'So I decided to unlearn everything in order to let the daily experience of Andean life teach me where to go.'

As he listened to his Quechua neighbours and walked with them through their fields, Gomel Apaza became aware that much of what he needed to know to improve crop yields was present in their ancient culture. This belief was confirmed when he gave courses for Chuyma Aru, an indigenous organization near Puno, and realized that agriculture could be based on local knowledge. In 1995, in his home village of Pucará, he launched the Asociación Savia Andina Pucará to promote the cultivation of a wider variety of potatoes and other native plants.

For a decade, Gomel and his neighbours demonstrated that diversification of seeds and tubers, along with traditional methods of preparing the soil, enhanced crop and grassland yields. Although the region is economically impoverished, he showed that by reviving the diversity of their natural heritage, rather than resorting to imported chemicals and technology, all farmers could produce enough to feed their families.

Gomel, aged 37, has been selected as an Associate Laureate in the Rolex Awards for an ambitious project to encourage more than 500 families in the areas around Orurillo and Pucará to broaden the genetic variety of their crops. More than 100 village gatherings and other public events will be held for Gomel and his team to share information with farmers. Encouraging agrodiversity, Gomel explains, is a key to combating hunger: 'A diversity of plants has more possibilities of surviving adverse environmental conditions. We have very extreme weather in the highlands, and if it gets very cold and you only have one

Returning to his roots, Zenón Gomel Apaza realized that the ways of his ancestors produce more and better potatoes and other vegetables than the ways he learned at university. The Associate Laureate says diversity is the key to better yields in the rugged Altiplano, his family home for generations. Here, where the potato was first domesticated, are found the greatest number of wild species.

'Zenón Porfidio Gomel Apaza brings strong personal qualities to his work: commitment to the good of local communities and belief in the importance of preserving biodiversity.'

Heliodoro Díaz Cisneros former programme director of the Kellogg Foundation for Latin America and the Caribbean

type of potato, you could lose everything. But when there is diversity, some types may die, but not others.'

For example, while most varieties of potato grown around the world belong to a single species (*Solanum tuberosum*), in the Andes – the potato's birthplace – about ten different *Solanum* species are cultivated, and wild potatoes provide over 200 additional species. About 5,000 potato varieties have been identified by the Peru-based International Potato Center, and scientists say no other major food crop enjoys such genetic diversity. Behind the sturdy tuber's multiplicity lies the ingenuity of Andean farmers, whose intimate knowledge of mountain agriculture has constantly produced new diversity, allowing them to plant potatoes chosen for the soil's quality, temperature, inclination, orientation and exposure. For more than 10,000 years, they enriched their genetic stock by swapping seeds. Yet the same farmers who used to harvest several dozen varieties have for years been pressured by agricultural technicians and agribusiness to reduce the types they cultivate. The 'Green Revolution' of the 1960s, with its focus on pesticides, machines and high-yield hybrids, increased the vulnerability of Andean people by narrowing the genetic base of once self-sufficient farming communities.

For many years, Gomel Apaza's father grew dozens of potato varieties, some for baking, some for soup, still others for medicinal purposes. But, Gomel says, he embraced the 'fad of modern technology and the Green Revolution' and cut back, cultivating only five

Ritual is important to the Quechua of Peru, whether raising root crops for food or alpaca for wool. As a sign of respect to Mother Earth, officials place their hats on the ground and toss a libation on the field before it is planted. An *ayllu*, an extended family group, farm together, using the *chakitaqlla*, the Andean foot plough, to work the soil. The old ways bring a happy harvest to farmers near Orurillo Laguna (overleaf).

Red potato, blue potato – the world's most widely grown tuber comes in many sizes, shapes and colours. The Quechua also grow varieties of quinoa, a prize grain of the Altiplano called the 'mother of seeds' by the Inca. To teach the next generation when to plant and harvest, a teacher drew up an agricultural calendar according to Altiplano traditions. Very cold nights are needed to make *chunyos*, potatoes that are 'freeze-dried' for 15 days or more. These women turn the crop daily.

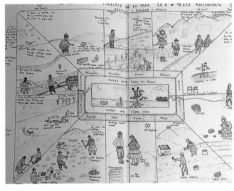

varieties. Gomel is undoing the damage of recent years by diversifying what is cultivated, as well as by setting aside natural reserves containing plant varieties, many of which may one day cure diseases.

Like the potato, other tubers such as ocas, izanos and ollucos, as well as grains such as quinoa and cañihua, are also being researched in the Pucará and Orurillo regions, and the project will protect 22 hectares of microhabitats of native plants. Hillsides eroded by inappropriate agricultural techniques will be recovered for use in traditional, environmentally sustainable ways. Gomel is helping to restart regional fairs where farmers gather to exchange seeds and discuss their crops.

Besides promoting agrobiodiversity in farmers' gatherings, Gomel is extending his advocacy into other public realms, using radio spots and working with educational institutions to promote agriculture suitable for the Andes. He is pushing primary schools to expand class curricula and to synchronize their school calendars with the long-established agricultural calendar.

Gomel Apaza's work has attracted support from organizations that are enthusiastic about his use of traditional methods to solve pressing problems. Heliodoro Díaz Cisneros, a former director for Latin American and Caribbean programmes of the W.K. Kellogg Foundation, which supported Gomel's project in Pucará, says that by encouraging farmers to rely on the knowledge inherited from their ancestors, Gomel 'encourages cooperation with neighbours and respect for the environment'.

Embracing the lessons of the past will, Gomel Apaza is convinced, produce more than just more potatoes – it will transform how communities are governed, as neighbours relearn the respect for the earth and each other that local culture emphasizes. 'Andean agriculture is not a substantial modification of the landscape, but rather a kind of beautification of it,' Gomel Apaza explains, adding that, for him, 'the relation between people and nature exists within a framework – based in caring and ritual – of feeling that you belong to all that exists.'

'Growing rare varieties of potatoes on the Peruvian Altiplano seems simple and humble at first glance, but behind it is a salute to and fight for the preservation and value of indigenous knowledge and practices. The spirit of the project reminds me of no less than the work of Rigoberta Menchú.'

Sir Magdi Yacoub heart surgeon and professor of cardiothoracic surgery at Imperial College, London

High-altitude tubers include varieties of ocas and ollucos, second only to the potato as highly valued food crops. Also on display at a village market, a large array of *chunyos* lures Gomel Apaza. His project is showing that traditional methods are best for his people. 'Diversity is a guarantee of food security,' he says.

Protect the Snow Leopard by Insuring its Prey

By Bob Guthrie

The snow leopard is an elusive, endangered feline – but it is also a powerful predator. Stalking livestock at high altitudes, it is the main foe of herders protecting their flocks. But, by combining ecotourism and insurance for livestock, economist-turned-conservationist Shafqat Hussain has made it possible to insure domestic herds, thus pacifying local herdsmen. Convinced that man and beast can share the same territory, he is now extending his scheme.

♛ Shafqat Hussain Pakistan

From time immemorial, shepherds in the Himalayan mountains of Baltistan, in northern Pakistan, have reviled the snow leopard as much as their Western counterparts once hated wolves. About half of the Balti economy comes from animal husbandry, particularly domesticated goats that are preyed upon by the snow leopard, largely because its traditional wild prey – the ibex and markhor – are disappearing. So local herders do not hesitate to kill the snow leopard, which is also at risk from the illegal trade in its highly prized pelt.

In an impressive example of lateral thinking, Shafqat Hussain, who originally trained as an economist, created Project Snow Leopard (PSL) in 1998 to deal with the threat to the snow leopard in Baltistan. This non-profit conservation programme combines ecotourism and low-cost insurance, protecting herders against attacks by the leopards on their livestock. The plan is helping local people realize that one cat alive in the surrounding bush is worth more to them than several killed for the fur trade.

Hussain, aged 37, describes the snow leopard as 'a marvel of nature's perfection' and explains that, sitting at the top of the food chain, this animal plays a key role in maintaining the mountain ecosystem. Dr Ma Ming, of the Snow Leopard Trust in Xinjiang, China, calls it an 'umbrella species': protecting it ensures its habitat and many other local species are also preserved.

Wonderfully adapted for the extreme weather and rocky terrain, the snow leopard roams wild at altitudes up to 5,500 metres in the Himalayan peaks. Furry feet help it stay on top of the snow by acting as natural snowshoes. This rare creature hunts alone for wild and domesticated goats and other prey, which it pounces upon from up to 15 metres away. With a total population estimated at between 4,000 and 7,000 scattered across the Himalayas, including fewer than 150 in Baltistan, the snow leopard is listed as 'endangered' on the IUCN Red List of Threatened Animals. This elusive relative of the tiger and more familiar African leopard is one of the least photographed, but most photogenic of big cats, with its metre-long tail and handsome dappled coat.

Man with a mission, Associate Laureate Shafqat Hussain used his economic background to devise a way to convince herders to seek compensation, not revenge, when their goats, like this little one, are preyed upon by snow leopards. His Full Moon Night Trekking company guides ecotourists to wildlife in the mountains of Baltistan. Profits support Project Snow Leopard and its insurance plan.

'Combining economics and conservation to save the snow leopard shows imagination, clear thinking and deep concern for humans and animals in the Himalayas.'

Luis Rojas Marcos professor of psychiatry at New York University School of Medicine

Men from a Baltistan village meet Hussain to establish a committee to oversee and pay claims for loss of livestock. In 2004, 18 local goats and sheep were killed by a snow leopard trapped in a pen. Despite the predation, the villagers decided to free the animal which one man said 'captures the spirit of the wild'. The ibex (below) is the snow leopard's favourite wild prey.

The insurance scheme set up by Hussain compensates villagers for every goat killed by the predators, which effectively deters the villagers from killing the offending cat or any other suspect. The annual premium paid is one per cent of the value of one goat, with each herder paying according to the number of goats he owns. This covers about half of all claims. The other half comes from Full Moon Night Trekking, the ecotourism agency Hussain founded, which advertises the snow leopard as its chief attraction. 'People who find pleasure in the idea of the snow leopard surviving in the wild should be willing to pay for this pleasure, and this payment helps offset the losses to farmers for having the cat around,' he says. To succeed, both complementary programmes must be profitable, another reason for locals to protect the animal at the heart of the project. A key aspect of the scheme's success is the fact that local people participate at every level. Village committees collect premiums, pay claims and act as the scheme's financial watchdog. Villagers control the income from Full Moon, using surplus profits for community projects, like making wells for drinking water. Full Moon also employs two villagers as guides.

Until recently, Project Snow Leopard covered a relatively small area of 170 square kilometres embracing the environs of the village of Skoyo, which has 260 inhabitants, and other nearby settlements. With support from his Rolex Award, Hussain is now extending his project to more Balti villages near Skoyo and K2, the world's second-highest mountain. He also hopes to attract more ecotourists, many of whom are staying away because of the 2005 earthquake and bad publicity about Pakistan in the wake of 9/11. Hussain points out that Baltistan is very much associated with adventure tourism, but he wants it to be known for ecotourism. 'Things can change and other trekking companies now mention wildlife in their brochures because Full Moon started to do it,' he says. 'In Nepal [also home to snow leopards], they have about 200,000 visitors per year; here in Baltistan, we only have about 5,000.'

'Shafqat Hussain's work is groundbreaking…effective, thoroughly designed and appropriately implemented.'

Rodney Jackson founder-director of the Snow Leopard Conservancy

From valley floor where flocks graze to mountain ridges where ibex roam, the realm of the snow leopard ranges from 3,600 to 5,500 metres high. Here in Baltistan, Hussain estimates there are less than 150 of the endangered cats. Ever elusive, the cats can be captured by remote cameras. Triggered by a photocell, a camera records one feasting on a domestic goat.

The added funding will also allow him to build better fences to protect livestock and to update the counting of leopards, mainly by automatic unmanned cameras dotted across the mountain landscape.

Hussain's broader vision is to demonstrate that human villagers and feline predators can live side by side. By involving local people, he is gradually convincing the villagers that man and beast can profitably coexist. He sees it as 'sadly ironic' that in many places there is more concern for endangered biodiversity than for humans. 'No matter how charismatic an animal is, its survival should not come at the cost of poor human farmers,' he says. But he adds that he is 'only one of many who are trying to make a difference for snow leopards and herders', pointing out that his project would not survive without local colleagues who run the scheme when he is in London and at Yale University writing his Ph.D. thesis on the relationship between human societies and the natural environment in the mountains of northern Pakistan.

His innovative idea of enlisting local support and willingness to challenge widely accepted ideas have won him many admirers, including Rodney Jackson, the renowned snow leopard authority who won a Rolex Award in 1981 for his work to save them. 'Hussain's work is groundbreaking, especially since he brings the social and ecological sciences together,' Jackson says. 'My impression is that Hussain's original programme is more effective [than similar ones inspired by his scheme], thoroughly designed and appropriately implemented.'

Dramatic valleys mark the route to Skardu, capital of Baltistan. Hussain and his associates meet at Full Moon Night Trekking headquarters in Skardu before hiking to the heights above the Skoyo valley. Political conflicts in the region have meant less tourist revenue to fund the insurance project, but, by including a full-moon night in most mountain treks to glimpse snow leopards, the company hopes to boost numbers.

Internet Connects Those Who Whistle Language

By Manuela Palma de Figueiredo

Language is undoubtedly one of the most amazing creations of the human mind. But today over 50 per cent of languages are in danger of disappearing, including some that have developed a fascinating form: whistled and drummed languages. A young French bio-acoustician and linguist, Julien Meyer, wants to revitalize these little-known ways of communicating with the help of the people in many remote parts of the world who use them.

Julien Meyer France

Suddenly, in the constant rustling of the Thai jungle, a clear, strong whistle cuts through the air. Meaningless to the uninitiated, this melodious phrase, resembling birdsong, carries precise information: hidden in the dense tropical vegetation, a hunter from the Hmong people is sending a long-distance message to his fellow-hunters about their plan for trapping a wild boar they have been tracking for hours.

This event, like something from the ancient past, is by no means confined to one isolated group. Unknown to most people, and marginalized by linguists, whistled languages have been used the world over for millennia, but are now threatened with extinction within this generation or the next. Passionately interested in languages and all modern forms of communication, Julien Meyer, a 30-year-old French bio-acoustician and linguist, refuses to simply do nothing while a part of the world's heritage is threatened by the movement of people from the countryside to the cities and by the emergence of new technology. Over the past 10 years, he has verified the existence of 34 whistled and drummed languages throughout the world, and devoted his skills and energy to studying, documenting and preserving a dozen of them. For his determination to safeguard a fast-disappearing, age-old practice, Julien Meyer has been selected as an Associate Laureate in the 2006 Rolex Awards.

Whistled languages communicate over distances like a mobile phone, but they are free and no technology is required. They faithfully transpose the grammar, syntax and, syllable by syllable, the vocabulary of the spoken languages they are based on, producing an accurate rhythmic and melodic copy of them. Other languages exist in drummed form, which is less precise and more repetitive, and is used more for making public announcements than for dialogue.

Whistled and drummed languages are used in Latin America, Europe, Africa, Asia and Oceania, in remote areas which have a very rich biodiversity. They overcome distance – up to 30 kilometres for talking drums – and cut through background noise, demonstrating the extraordinary adaptability of groups living in mountainous areas and dense forest, where communication is a constant challenge. 'Whistled and drummed speech unites humans and nature by means of

A polyglot himself, Julien Meyer is intent on saving the little-known forms of languages that are whistled or drummed. To date, the Associate Laureate has identified 34 such forms of speech. In northern Thailand a member of the Akha people (preceding pages) uses a leaf to whistle a message that will carry a far longer distance than speech in these rugged hills.

'Language is so built into the fabric of human beings that we often overlook its crucial role as a foundation of culture. With the cultural heritage of so many languages about to disappear, this project is highly relevant.'

Laretna Adishakti founder of the Center for Heritage Conservation, Yogyakarta

Chirping like a bird, Luis Morales Mendez is a *silbador*, an expert in the whistled form of Spanish used in Gomera, one of Spain's Canary Islands. At an indigenous peoples' conference in Pau, France, a whistler from Colombia compares notes with Meyer. A Mazatec from southern Mexico whistles to a bird in a painting by artist Fernando P. Alonso.

language,' Meyer explains. 'Sound needs the natural environment as a carrier to propagate it over a long distance. In addition, these communication methods are a unique source of information about their users' environment and social life.'

The first studies of whistled languages were carried out in 1950 by Professor René-Guy Busnel. This famous French scientist, now retired, was the first to study this form of language in terms of linguistics as well as acoustics. Since then, however, whistled speech has raised little interest among linguists, and almost half a century went by before Meyer took up the cause of this fascinating method of communication.

In 1997, while studying at the Ecole Supérieure d'Ingénieurs (school of higher engineering studies) in Marseille, France, Meyer dreamed of working on languages and being able to apply his technical knowledge to concrete cases. He stumbled on an article about Béarnais, an extinct whistled language from the Pyrenees Mountains in France. It was an eye-opener for Meyer. 'It struck me that whistled languages provided a natural link between telecommunication systems and human language,' he recalls. He immediately immersed himself in the literature about this unusual subject and began planning visits to regions where people use whistled and drummed languages. He taught himself the spoken languages of some of these regions (he now speaks six languages) and, once he had completed his diploma in bio-acoustics, he set about acquiring the linguistic skills necessary to study whistled and drummed languages. During this time he discovered Busnel's work, and was spellbound. Eventually they met, and from the first discussions, it was a meeting of like minds. 'Julien is the inheritor of my scientific past,' says Busnel, who was born in 1914.

Faithful to the pioneering thinking of the man he regards as his 'oldest friend', Julien Meyer was convinced that the key to understanding whistled languages lay in studying them acoustically as well as linguistically. In 2003, he travelled around the world, forging close links with whistling communities and master drummers in France, Spain, Mexico, Brazil, Peru, Colombia, the Republic of Vanuatu, Laos,

Thailand, Nepal, Turkey and Greece. During his travels he recorded about 30 hours of whistled and drummed languages for subsequent analysis using the most advanced acoustic techniques. The recordings also provided material for his doctoral thesis on the intelligibility of whistled languages, written in 2005 for the University of Lyon 2 and the National Centre for Scientific Research (CNRS), in France.

This was only the start, however, for the brilliant bio-acoustician who had an even bigger goal in mind: to preserve this priceless cultural heritage at risk of constant erosion by modern technology – which could, ironically, also hold the key to keeping this heritage alive. Meyer will now set up, with the funds from his Rolex Award, an interactive Internet site featuring recordings, photographs and documentation on whistled and drummed languages. This project, 'The World Whistles', will be undertaken with close cooperation with the people who use whistled and drummed languages. They will also contribute to the site and oversee the use of the data on it. 'By giving them the opportunity to take over modern technology for their own use, and to communicate with other whistling and drumming people whose existence they never even dreamed of, I'm hoping to revive their belief in their own culture. Whistled and drummed languages belong to the people who use them,' insists Meyer, for whom human beings are clearly more important than scientific results. 'Respect for our fellow man is the first condition in acquiring knowledge.'

' I am convinced that the breadth of Julien Meyer's research will change the way whistled languages will be talked about and known. The whistlers stand to gain tremendously from his work on their behalf.'

Colette Grinevald UNESCO consultant for endangered languages

'Yo canto, tu silbas' (I sing, you whistle) is the title of this Fernando P. Alonso painting. The whistled speech of the Mazatec people reflects the structure and grammar of their spoken language; Juan Casimiro demonstrates his fluency. The success of an interactive website, set up by Meyer, will be ensured by the participation of those who use whistled and drummed languages.

Explore by Kayak the Labyrinths of Western Patagonia

By Paul Jeffrey

Eager to protect the dramatic landscapes of western Patagonia, Cristian Donoso will lead an expedition by kayak to this region, one of the most inhospitable places on earth. Spending five months navigating open seas and fjords and pulling their kayaks across glaciers, Donoso and his team will face daunting physical and mental challenges as they gather information that will inform Chile and the world about this little-known area.

Cristian Donoso Chile

With its labyrinth of rocky islands, serpentine channels and icy fjords, western Patagonia, in southern Chile, is one of the least-explored areas on earth, with annual rainfall reaching up to eight metres and winds frequently rising to hurricane force. Nestled among glaciers that hug the slopes of steep Andean peaks and drenched by storms that blow out of the southern Pacific, the harsh region deters all but the hardiest explorers.

That has not stopped Cristian Donoso, a young Chilean lawyer who over the past 14 years has ventured almost 40 times into the region's most inaccessible corners. Just like the indigenous peoples who paddled their fragile canoes here for thousands of years before the arrival of Europeans, he often travels in a sea kayak, a shallow craft that allows him to manoeuvre around the narrowest fjords and discover their hidden beauty.

'In order to strengthen the protection of this territory, we have got to know what's there,' says Donoso, who reports that today most Chileans have little knowledge of it. Along with team member Richard Vercoe, a naturalist from the United States who has documented the impact of economic activities on Chile's environment, he warns that such ignorance makes it easier for those seeking commercial gain to exploit the region's natural resources – seafood, water, virgin forests – with little respect for its biodiversity. Selected as an Associate Laureate for his fearless commitment to exploration and for his plan to gather vital new knowledge of western Patagonia, Cristian Donoso believes that his next major expedition will ensure greater public awareness of the region.

With his team of three men and one woman, the 31-year-old explorer is planning an ambitious five-month Transpatagonia Expedition starting in September 2007. They will traverse 2,039 kilometres of the central part of western Patagonia on open sea, lakes and rivers, as well as travelling overland for about 150 kilometres – including 22 kilometres atop glaciers, dragging their kayaks with provisions, weighing 200 kilograms each, behind them as sledges. The group will ascend unclimbed peaks and visit uncharted territories. Expedition

A passion for adventure and exploration brings Cristian Donoso to the wild side of Chile's Patagonian coast. In 2007–8 he and his four teammates plan a five-month expedition to probe the region's fjords, lakes and open seas by sea kayak, carrying out geological, ecological and historical surveys. The region is spectacular, but also fraught with danger from rugged terrain to violent weather.

'With profound respect for the significance of navigating through such pristine areas, Cristian Donoso wants to know these areas, love them and, by publicizing their millennia of history, preserve them for many more millennia.'

Mariela González professor of physical education at the University of Concepción

Starting point for a training mission, the town of Tortel, tucked in a Patagonian fjord, has no roads; wooden walkways link the houses. The team travelled to the region on the Carretera Austral, the main road to southern Chile's ice fields, largest in this hemisphere, after Antarctica. From Tortel, the team took a boat towards Glacier Montt, which is thinning rapidly.

members hope to encourage the region's small indigenous nomadic community of Kaweskars (or Alakalufs) to reclaim their ancestors' canoeing skills and host adventure tourists. Donoso, who plans to write a guidebook to the region, believes such sustainable economic ventures will help assure the region's protection.

Detailed plans for the odyssey include locations to camp each night and a system, designed by Donoso, that will allow the team to sleep suspended from the cliffs rising out of the frigid waters when no suitable campsite is available. The explorers will bring their own food, but supplies will be replenished twice by a boat from Puerto Edén, a small indigenous village where the Chilean Navy maintains a base. The trip will place great demands on the kayakers, who have begun an intensive physical and nutritional training programme and are making three-week training runs into the region.

The team will carry sophisticated first-aid equipment. In case of a serious accident or illness, the supply boat can come to their rescue – though during much of the trip it would take three days to reach them. To enhance understanding of the region's geological past, soil and rock samples will be collected, shipped out on the supply boat and analysed by university scientists. The explorers will also collect fossils and inspect geological evidence, including stalagmites in caves on Madre de Dios Island, showing how the climate has changed over time. The team's scientific investigator, Chilean geologist Rodrigo Fernández, participated in a landmark expedition in 2000 to Madre de Dios for which the leader, Jean-François Pernette, won a Rolex Award in 1998.

Scholars of the region's human history eagerly await the expedition's reports on the remains of fishing and hunting camps that belonged to the Kaweskars, who travelled the region for more than 4,000 years. Team member Kai Salas, a French archaeologist, will carefully document and site the settlements using GPS units.

A famous incident, the 1741 sinking of the English frigate *Wager* on the north coast of the Guayaneco Archipelago, will come alive again when the explorers dive into the sea to seek the wreck's exact location.

The main mode of travel for the expedition, sea kayaks can reach places inaccessible to larger boats – and they are easily transported. On this training mission, one of several planned before the expedition launch in September 2007, ice floes barred the way to Glacier Montt, but the return journey took them to the pristine beauty of Lake General Carrera (overleaf).

They will then seek to trace the route narrated in the journal of John Byron, who survived the shipwreck thanks to assistance from two indigenous groups who spirited him and three other survivors through the treacherous waters in their canoes.

Throughout the journey, a website will track the expedition's progress, with the explorers providing updates by satellite phone. One team member will produce a documentary video for broadcast on television in Chile in 2008.

According to team member Mariela González, a professor of physical education at the University of Concepción and a skilled kayaker, Donoso's comprehensive vision of Patagonia helped convince her to join the group. 'He has a deep commitment to showing Patagonia from a wide perspective, combining the different worlds of science, sport, history, photography, ecology and interviews to tell the stories of people who live there,' she says. 'With profound respect for the significance of navigating through such pristine areas, he wants to know these areas, love them, and, by publicizing their millennia of history, preserve them for many more millennia.'

'Cristian Donoso's expeditions to the most inhospitable parts of western Patagonia demonstrate his adventurous spirit and the commitment, endurance and curiosity of a true explorer.'

Erling Kagge polar explorer and president of Kagge Forlag Publishing

The Rolex Awards

The Rolex Awards
2006 Selection Committee

The Laureates and Associate Laureates of the Rolex Awards for Enterprise are chosen by a panel of independent, voluntary experts, who have international stature and are respected as leaders in their fields. This group of renowned, multidisciplinary specialists faces the challenging task of assessing the best projects that have been submitted from around the world and selecting the winners. A new panel of experts is invited to judge each series. For the current series, a total of 1,671 entries were submitted from 117 countries. The Selection Committee met at Rolex headquarters in Geneva in April 2006, under the chairmanship of Rolex Chief Executive Officer Patrick Heiniger, for the judging process. Since 1976, 96 men and women from dozens of countries have served as judges of the Rolex Awards.

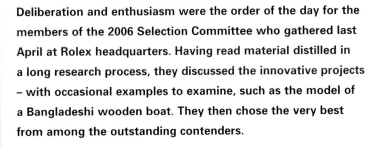

Deliberation and enthusiasm were the order of the day for the members of the 2006 Selection Committee who gathered last April at Rolex headquarters. Having read material distilled in a long research process, they discussed the innovative projects – with occasional examples to examine, such as the model of a Bangladeshi wooden boat. They then chose the very best from among the outstanding contenders.

'Choosing the winners was most enjoyable, but also challenging, because there are so many very good projects. It is also humbling to see how much human ingenuity is out there. Generally you only hear about bad things, but here are fantastic things being done by enterprising people.'

Sir Magdi Yacoub Member of 2006 Rolex Awards Selection Committee

2006 Selection Committee

Mr Patrick Heiniger – Switzerland
Chairman of the Selection Committee
of the Rolex Awards for Enterprise
Chief Executive Officer of Rolex SA

Dr Laretna Adishakti – Indonesia
Architect and engineer, founder of
the Center for Heritage Conservation,
Yogyakarta

Prof. Denise Bradley – Australia
Educator, vice chancellor of the
University of South Australia

Ms Motoko Ishii – Japan
Lighting designer, president of
Motoko Ishii Lighting Design Inc.

Mr Erling Kagge – Norway
Polar explorer, mountaineer and lecturer,
president of Kagge Forlag Publishing

Prof. Tommy Koh – Singapore
Diplomat and patron of the arts

Mr William K. Reilly – United States
Lawyer, president and CEO of Aqua
International Partners, former administrator
of the U.S. Environmental Protection Agency

Dr Luis Rojas Marcos – Spain and
United States
Psychiatrist, chairman of the NY Public
Health and Hospital Corporation

Mr Mark Shuttleworth – South Africa
Entrepreneur, technology specialist
and philanthropist, chairman of the
Shuttleworth Foundation

Sir Magdi Yacoub – United Kingdom
Heart surgeon, professor of Cardiothoracic
Surgery at Imperial College, London

Dr Laretna Adishakti
Indonesia

Professor Denise Bradley
Australia

'Heritage conservation is an integral part of my life,' says Laretna Adishakti. 'My spirit is there.' For the much respected Indonesian architect, heritage is any cultural tradition that is a legacy of our collective past and reflects our living environment, including art and culture, oral tradition, flora and fauna, artefacts, buildings and monuments – the entire 'cultural landscape'.

Adishakti, popularly known as Sita, was born into a family of conservationists who stimulated her interest in Indonesia's traditions and inspired her studies: she received an engineering degree from Gadjah Mada University in her native Yogyakarta (1982), a Master's in architecture from the University of Wisconsin (USA, 1988) and a doctorate in engineering from Japan's Kyoto University (1997).

Employed as a lecturer in several disciplines at Gadjah Mada University since 1983, Adishakti has, in her words, 'spread as many heritage "viruses" as possible' to encourage Indonesians to conserve their rich heritage. As head of the country's leading heritage conservation organizations, she is spearheading a campaign to protect the city of Yogyakarta, the cultural heart of Java; restoring the endangered Tamansari Water Castle, once a sacred site for a sultan and his entourage; and studying the cultural landscape of Borobudur Temple and its environment to ensure its proper heritage management. Adishakti's influence also extends internationally through her lectures and as a member of UNESCO-ICCROM's Asian Academy for Heritage Management and of the International Council on Monuments and Sites. Complementing her architecture and her many heritage projects, Laretna Adishakti is a devotee of the arts, including Indonesian flower arranging.

Denise Bradley, vice chancellor and president of the University of South Australia (UniSA), the state's largest university, is renowned for her unflagging efforts to raise the standard of higher education in Australia by continuously meeting difficult challenges. Named one of the *Australian Financial Review*'s top 25 'True Leaders' of 2003, she attributes her success to finding solutions rather than identifying problems. 'When something different happens, I think it's exciting and not a threat,' says Bradley, who tries to instil this principle in both students and her academic team.

Professor Bradley began work as a secondary school teacher in the early 1960s and soon became disillusioned with the barriers faced by women teachers. In the mid-1970s, she set out to improve women's education and employment opportunities, and, by the end of the decade, had been appointed a government advisor on women's education. In 1997, she became head of UniSA, the third woman ever selected as a president of an Australian university.

Over the years, Bradley has served on national and international bodies, advising governments on educational policy. 'We must endeavour to make education truly open to all,' she asserts. In addition to her educational activities, Bradley is a key promoter of South Australia where she acts as a business ambassador and is involved in the Adelaide Festival, the state's main cultural event. 'Great leaders have a broad knowledge of far more than what's happening in their own industries,' she says.

For services to her country, Professor Bradley was made an Officer of the Order of Australia in 1995 and was awarded the Centenary Medal in 2003.

Motoko Ishii

Japan

For Motoko Ishii, 'light is life', and the internationally acclaimed light-ing designer has spent her life fulfilling a wish 'to show the beauty of light to as many people as possible'. A pioneer in her field, she has extended the boundaries of lighting design in urban, architectural, landmark and environmental illumination.

A childhood interest in things mechanical and a desire to become an industrial designer led Ishii to study product design at Tokyo National University of Fine Arts and Music and to subsequent work in Finland and Germany. In 1968, she established Motoko Ishii Lighting Design, which, along with sister company Light Creation, has imple-mented hundreds of innovative projects worldwide. Among her many outstanding and varied commissions have been Expo '70 in Osaka; the Akashi Kaikyo Bridge, the world's longest suspension bridge; the Shi-rakawago historic village, a World Heritage Site; Tokyo Station; Tokyo Tower; laser-art performances in Vienna and Brussels; and a produc-tion of Puccini's *Madame Butterfly*.

Harmony with nature is fundamental to Ishii's environmentally friendly lighting systems, often designed to revitalize structures and enhance heritage sites. Lighting should be considered from the stand-point of culture, not technology, says Ishii, who has received dozens of prestigious awards in Japan and abroad, such as the Shiju-hosho (Medal of Honour with Purple Ribbon) from the Japanese government. Her work, including lighting fixtures she personally designed, is pre-sented in numerous publications, the latest, *Lighting Horizons*, in 2001. Of her future, Ishii says: 'I want to explore some of the possibili-ties that new technologies create.'

Erling Kagge

Norway

Norway's most acclaimed living polar explorer and one of the greatest adventurers of our time, Erling Kagge was the first person to surmount the 'three poles' – North, South and the summit of Mt Everest. For the past 15 years, he has been setting new standards in exploration, living up to his motto: 'If you can dream it, you can do it.'

Skiing to the Earth's extremities, sailing the oceans, climbing mountains and reaching beyond himself have been Erling Kagge's goals since childhood, ambitions inspired by his boyhood heroes, fellow Norwegians Roald Amundsen and Thor Heyerdahl, as well as Albert Schweitzer. Kagge had already sailed across the Atlantic twice, around Cape Horn and to Antarctica and back by the time he received a law degree from the University of Oslo in 1989. He had also begun training for his record-breaking expeditions: in 1990, he became the first man – together with Børge Ousland – to reach the North Pole unsupported. Three years later he was the first to reach the South Pole, walking alone and unsupported for 50 days – an exploit featured on the cover of *Time* magazine.

For two years during this period, Kagge worked as a lawyer for industrial giant Norsk Hydro. In 1996, after a year's sabbatical reading philosophy at Cambridge University, he founded what is today one of Scandinavia's most profitable publishing houses. In addition to running his business and collecting contemporary art and Russian icons, Kagge lectures frequently to geographical societies and business organizations. His three books on his polar expeditions have sold more than 60,000 copies. 'Not everyone can do what I did, but I recommend that each person finds his own South Pole,' says Kagge.

Professor Tommy Koh
Singapore

'I have always been driven by the wish to build a better world,' says Professor Tommy Koh. For nearly four decades, the highly esteemed Singaporean has steadfastly served his country as a jurist, diplomat, environmentalist and vigorous champion of the arts. Formerly Singapore's ambassador to the United States and to the United Nations, the much-awarded Koh today serves as ambassador-at-large at the Ministry of Foreign Affairs, chairman of the Institute of Policy Studies and the National Heritage Board. His three jobs – in diplomacy, in a think-tank and as head of Singapore's museums – help him maintain a youthful and positive outlook on life.

Koh's concern for justice can be traced to his high-school days when he resolved 'to help give voice to the voiceless'. Educated in law at the University of Singapore, Harvard and Cambridge, he became a respected lawyer and law professor. In 1967, when Koh was only 30, Singapore's then Prime Minister, Lee Kuan Yew, appointed him permanent representative to the UN, marking the start of his distinguished diplomatic career. Among his many accomplishments, he played a key role in negotiating the landmark 1982 UN Convention on the Law of the Sea and spearheaded the 1992 UN Conference on Environment and Development that helped the Earth Summit in Rio de Janeiro reach consensus.

Koh describes his approach to diplomacy as pragmatic idealism, a concept set out in one of his several publications, *The Quest for World Order*. 'I always try to achieve a consensus and to bring along all the stakeholders,' he adds. 'This approach has served me well, in diplomacy as well as in life.'

William K. Reilly
United States

William K. Reilly, one of the most influential conservationists in the United States, has championed progressive environmental policies for 40 years. His skill in mediating opposing factions, coupled with his commitment to preserve our natural systems, has gone far to help ensure better use of the world's resources.

Reilly credits his father with instilling in him an interest in land, history and justice. 'Conserving, planning, developing land requires reconciling nature with culture. It is my passion,' he says. Equipped with a law degree from Harvard (1965) and a Master's in urban planning from Columbia University (1971), Reilly held a series of senior, environment-related positions. While president of the World Wildlife Fund-U.S. (WWF) from 1985 to 1989, he innovated 'debt for nature' deals. 'Conservation must work for the people,' he adds. 'If human needs are not met, the system of protection will not endure.'

In 1989, Reilly became head of the U.S. Environmental Protection Agency (EPA) where he administered 18,000 employees and an annual budget of $7 billion. Over four years, he strengthened the role of science at the EPA and initiated pollution reforms, including a strong new Clean Air Act, among other accomplishments. In 1992, he led the U.S. delegation at the Earth Summit in Rio de Janeiro.

Currently, as president and founding partner of Aqua International Partners, which finances water improvements in developing countries, Reilly fulfils his interest in worldwide environmental concerns. He also serves on the boards of three leading companies, as chairman of the WWF, a trustee of the National Geographic Society and as co-chair of the National Commission on Energy Policy.

Dr Luis Rojas Marcos
Spain and the United States

Mark Shuttleworth
South Africa

Dr Luis Rojas Marcos, professor of psychiatry at New York University School of Medicine, is celebrated for his seminal work in treating the urban homeless, mentally ill immigrants and violent youth. The visionary psychiatrist has gained respect in both the United States and his native Spain for his pragmatic approach to helping people cope with life's challenges. 'Life is change, change is life,' philosophizes Marcos who has dedicated his own life to unearthing the explanations for human behaviour.

After graduating from Seville University Medical School in 1968, Dr Marcos left Franco's Spain and emigrated to New York City where he furthered his medical studies and conducted groundbreaking research on how language barriers and cultural differences contribute to the misdiagnosis of psychiatric patients who speak little English. For 25 years, he helped shape public health policy in New York, earning accolades for his innovations as head of the city's Department of Mental Health Services and the Health and Hospitals Corporation.

As well as teaching psychiatry in New York, Dr Marcos remains firmly connected to his roots in Spain where he serves as a trustee of La Caixa Foundation, which deals with social and public health problems, and frequently lectures on key medical issues. He also writes about social and psychiatric topics in English and Spanish, including a regular column for Spain's *El País* newspaper.

Marcos admits that writing, a very 'rewarding act' that binds him to his Spanish contemporaries, is something he finds stimulating – along with playing the piano and guitar and competing in the New York Marathon.

'Each of us has a dream, something we want to fulfil in life,' says technology specialist and philanthropist Mark Shuttleworth. The South African entrepreneur is renowned at home and abroad for seeking new opportunities in global technology. Investing in innovative breakthroughs, Shuttleworth's venture-capital firm HBD (Here Be Dragons, a phrase purportedly used to describe uncharted territory on early maps) exemplifies his enterprising spirit.

Shuttleworth became fascinated with technology as a child when he discovered computer games, a 'vice', he admits, that continues today. In 1995, during his final year at the University of Cape Town, he founded Thawte Consulting which quickly became the leading Internet security company for electronic commerce outside the U.S. The proceeds from the company's sale in 1999 have allowed him to fund several non-profit organizations, including bridges.org, whose mission is to help span the technology gap between Africa and the developed world, and the Shuttleworth Foundation, dedicated to improving education in Africa. Education helps young people to recognize that anything is possible, says Shuttleworth.

Shuttleworth's own long-held ambition was realized in 2002 when, as a crew member of Russia's Soyuz TM-34 spacecraft, he became the first African in space. The scientific experiments that he conducted on board the International Space Station form the basis of his Hip2BSquare roadshow, which aims to make both mathematics and science attractive to pupils and reinforces the message that 'knowledge is power'.

Professor Sir Magdi Habib Yacoub
United Kingdom

'I consider myself privileged to be able to serve science and medicine in a global fashion, because science and medicine know no boundaries,' says Professor Sir Magdi Habib Yacoub. A recognized leader in the field of heart and lung transplants – he has performed over 2,600 transplants, more than anyone else – the Egyptian-born heart specialist is hailed as one of the world's most respected cardiac surgeons for his pioneering techniques.

Influenced by his surgeon father and by the death of an aunt from heart disease in her 20s, Magdi Yacoub decided as a boy to become a heart surgeon. He entered the Cairo University College of Medicine at the age of 15 on a full scholarship and, following qualification in 1957, trained in the U.S. and Europe. In 1962, he moved to England where, over the next four decades, he held senior positions at leading London hospitals and carried out 20,000 open-heart operations. In 1986, he was appointed as the first British Heart Foundation Professor of Cardiothoracic Surgery, a position he still holds.

Today, Sir Magdi, the author of 800 scientific articles, continues his groundbreaking work as research director of the Magdi Yacoub Institute and through his international charity, Chain of Hope, which cares for poor children from war-ravaged countries. He also returns regularly to Egypt to treat children with heart conditions free of charge.

Sir Magdi, who was knighted in 1991 and elected to the Royal Society in 1998, views international healthcare delivery as a top priority along with his research. The dedicated surgeon still performs six or seven operations a week and, in his spare time, raises orchids and listens to Bach, an abiding passion.

♔ The Rolex Awards in Brief

'Over three decades, we have helped advance the exceptional work of scores of individuals who are quietly changing the world and making it a better place to live.'

Patrick Heiniger Chief Executive Officer of Rolex SA and Chairman of the Rolex Awards Selection Committee

History A Tradition of Enterprise

What, some could ask, do watchmaking and an awards programme supporting science, technology, exploration, the environment and cultural heritage have in common? The answer is simple: enterprise. Enterprise, in this context, is an undertaking marked by boldness and initiative. It is the inspiration upon which Hans Wilsdorf founded Rolex a century ago and set it on a course of unsurpassed innovation in creating the world's finest timepieces.

The late André J. Heiniger, former chairman of Rolex, exemplified this enterprising spirit, spearheading the company's outstanding growth over 50 years. He initiated the Rolex Awards for Enterprise in 1976 to commemorate the 50th anniversary of the company's greatest technical achievement, the waterproof Oyster chronometer, and to encourage human endeavour.

Thus, in keeping with the company's pioneering history, Rolex has assisted scores of Rolex Award winners in their efforts to bring groundbreaking projects to fruition. These bold and tenacious women and men have helped to make our planet a better place to live.

'I certainly believe that individuals can make, have made, and will continue to make huge contributions to knowledge in today's world.'

Gilbert M. Grosvenor

Chairman of the National Geographic Society Member of the 2000 Rolex Awards Selection Committee

Objective Supporting Pioneering Individuals

Rolex created the Awards for Enterprise to provide visionary men and women worldwide with the financial support and recognition needed to carry out innovative projects. The Awards are presented every two years. Although the winning projects differ greatly in subject matter and approach, they are similar in their intent to improve our planet and the human condition.

Unlike other programmes that reward past achievements, the Rolex Awards identify and assist individuals embarking on new ventures or completing ongoing projects. Typically working on their own and outside the mainstream, these adventurous and dedicated people often do not have access to traditional funding sources.

In supporting these pioneering concepts, the Rolex Awards play a unique role in nurturing the spirit of enterprise around the globe.

Areas of Recognition Advancing Knowledge and Well-Being

The categories of the Rolex Awards span five major areas of recognition:

Science and Medicine

projects in the natural or physical sciences that contribute to human health and welfare

Technology and Innovation

inventions, new devices or processes in the applied sciences that significantly contribute to society and to the world

Exploration and Discovery

expeditions, journeys or ventures that inspire our imagination, expand our knowledge of the world or shed new light on our planet

The Environment

projects that protect, preserve or improve our natural and physical surroundings

Cultural Heritage

projects that conserve, safeguard or contribute to our common historical, cultural or artistic heritage

These fields of endeavour are broadly interpreted to include almost any undertaking – from a technological breakthrough to a journey of discovery or a simple solution to a long-standing problem – as long as it ultimately benefits the world or contributes to our understanding of it.

Eligibility Individuals from all Walks of Life

The Rolex Awards are notable for the diversity of the individuals they support. Anyone of any age, nationality or background can apply. Winners come from every corner of the globe and have ranged from an engineer in Senegal to a palaeontologist in Canada.

Among the many recipients are an accountant who is today an authority on wildlife care, a taxi driver who has become a renowned beetle expert, a former teacher and carpenter who runs the world's biggest bicycle recycling operation, a winemaker who is also a world-class speleologist, and an amateur science writer, who, thanks to the Awards, is now a celebrated scientist in his own right.

In contrast to most other awards programmes that depend on third-party nominations, Rolex Award entrants put themselves forward. Applicants must show that they have turned an original idea into a concrete working project and how, through initiative and drive, they intend to execute their scheme and attain their stated goal.

'Human beings are very ingenious and there seems to be no end to the exciting projects that can be produced in every field of endeavour.'

Sir Edmund Hillary
Mountaineer and explorer
Member of the 1993 Rolex Awards
Selection Committee

'It is indeed the unreasonable, the unconventional individuals who change the world time and time again, who have reinforced my faith in the invincible spirit of mankind.'

The late André J. Heiniger
Founder of the Rolex Awards for Enterprise

Prizes Funding and Worldwide Recognition

Five Laureates – those who have demonstrated the most remarkable spirit of enterprise – are selected from thousands of applicants for each biennial Award series. The 2006 Laureates each received US$100,000 and a solid gold Rolex chronometer. They were honoured for their outstanding projects at an official awards ceremony.

The five Associate Laureates selected in 2006 each received $50,000 and a steel-and-gold Rolex chronometer at ceremonies in their home countries or regions. Award recipients must use their monetary prizes to implement or complete their innovative projects. While welcoming this often vital funding, many winners cite the global recognition and publicity they receive as the greatest benefit. Such international exposure helps to validate their work among peers and opinion-formers at home and abroad.

Selection Criteria Originality, Feasibility, Impact

Four main criteria are used to select the winning projects.

In judging each entry, the Selection Committee determines whether the project is feasible – can it be carried out successfully?

Is it an original concept, breaking new ground in a creative and innovative manner?

Does it positively impact on the surrounding community and the world at large? Do many people stand to benefit?

Above all, the judging panel considers whether candidates demonstrate a spirit of enterprise. Have they approached a unique undertaking with determination, tenacity and boldness? Have they had to face and overcome the most challenging odds?

Importantly, the judges also consider what effect a Rolex Award will have. To what extent will the Award allow recipients to implement or complete their projects? Will the funding and global recognition engender further support?

'The greatest adventure is still the voyage of the human spirit.'

Roberta Bondar
Neurologist and astronaut
Member of the 1998 Rolex Awards
Selection Committee

'I was really struck by the fact that while the judges bring different types of expertise to the table, nobody pushed a particular agenda, they advanced ideas, they advanced knowledge. We came together with debate certainly, but without dissension.'

Kathryn S. Fuller
Lawyer and environmentalist
Member of the 2004 Rolex Awards
Selection Committee

Judges Experts and Innovators

An independent jury made up of experts from a variety of disciplines and countries chooses the winners of the Rolex Awards. Members of the Committee, who serve on a voluntary basis, clearly embody the spirit of enterprise that the Awards seek to encourage and promote.

In the past, architects, anthropologists, chemists, explorers, surgeons, physicists, environmentalists, sociologists, engineers, writers, astronauts, archaeologists, oceanographers and mountaineers have acted as judges. Members of past Selection Committees include Sir Edmund Hillary, who climbed Mount Everest in 1953; Dr Leo Esaki, Nobel Prize physicist; William Graves, former editor-in-chief of *National Geographic* magazine; Roberta Bondar, neurologist and astronaut; Paul-Emile Victor, polar explorer; and Haroun Tazieff, volcanologist.

The Secretariat of the Rolex Awards for Enterprise is located at the headquarters of Rolex SA, Geneva, Switzerland.

P.O. Box 1311
1211 Geneva 26
Switzerland
Email: rae@rolex.com
Website: www.rolexawards.com

Head of the Rolex Awards Secretariat: Rebecca Irvin
Editors: Edmund Doogue and Joëlle Martin-Achard
Photo editor: Laura Bucciarelli
Photo assistant: Stéphanie Paccard
Caption writer: Elizabeth Moize
Editorial assistant: Regina Rueger Surur
Translator: Veronica Kelly
Scientific advisors: Quentin Deville, Barbara Geary and
    Francesco Raeli
Editorial consultant: Karen de Leschery

First published in the United Kingdom in 2006 by
Thames & Hudson Ltd, 181A High Holborn, London WC1V 7QX

www.thamesandhudson.com

British Library Cataloguing-in-Publication Data
A catalogue record for this book is available from the British Library

ISBN-13: 978-0-500-51337-8
ISBN-10: 0-500-51337-6

Design and layout: Grade Design Consultants, London
Printed by Amilcare Pizzi in Italy

Illustration Credits

Photographers for the Rolex Awards (T-top, B-bottom, L-left, R-right):
Kurt Amsler: 2–3, 4 (2T, 2B); 22–27; 29–35
Jacques Bélat: 5 (4B); 112–114; 116; 140
Tomas Bertelsen: 4 (1T, 1B); 5 (4T); 10–17; 19–21; 110–111; 138TR; 142
Thierry Grobet: 5 (3B); 102–104; 106–109
Didier Jordan: 8; 9; 128–137
Marc Latzel: 4 (3T, 3B, 5B); 36–49; 66; 72T–75; 143
Xavier Lecoultre: 4 (4T, 4B); 5 (2T, 2B); 50–62; 88–99; 141T
Heine Pedersen: 5 (1T, 1B); 76–87; 138L, 138BR; 141B
Stefan Walter: 5 (5T, 5B); 118–127

Sutter Agency: NHPA/Kitchin T. & Hurst V.: 100–101; NHPA/Kirchner R.: 5 (3T); Wildlife Pictures/Pölking F.: 105

Laureates' map (6–7): Léonie Schlosser

Rolex wishes to thank the following people for use of their images:
Shafqat Hussain and Hushe Village Conservation Committee: 107TR
Julien Meyer and artist Fernando P. Alonso: 115, 117
Brad Norman/ECOCEAN: 28
Pilai Poonswad/photo by Tsuji A.: 18
Michael Proudfoot: 71
Rory Wilson: 68; 69T; 70; 72B; photo by Zimmer I.: 64–65; photo by Gleiss A.: 69BL; photo by Liebsch N.: 69BR